Preface

Shiftwork, including rotating and fixed shifts, is different from any other job situation. People who work on rotating shifts or on fixed shifts that involve night, evening or early-morning work hours face unique demands on their bodies and their time. Shiftwork also can mean changes in lifestyle for whole families, especially when weekend work is involved.

This **Introductory Guide** contains ideas that will help you and your family handle some of the changes that shiftwork can create in your life. The first section, Keys To Shiftwork Success, identifies those field-tested factors that lead to a healthy, productive and successful life on shiftwork. The second section of the **Guide**, The Body Clock, explains why you, your body, and your family and friends are challenged by your work schedule. Each of the remaining five sections — Sleep, Eating, Fatigue, On-Shift Alertness and Family And Social Time — then talk about particular challenges you might face, and suggest ways you can resolve specific problems or even prevent these challenges from occurring in the first place.

As you read the **Introductory Guide** and try the ideas presented here, we encourage you to share this new information with your family and friends. It will help them understand what shiftwork is all about. And their understanding and support are among the most important factors in helping you to cope well with shiftwork.

As you and your family make the changes you know are necessary, you will find that your work and leisure time are more productive and that you are both healthier and happier. We applaud you for taking this journey towards a successful life on shiftwork.

Contents

Keys To Shiftwork Success

Why Is Shiftwork So Common Today?

Today, an ever-growing number of people worldwide work some type of evening, night, early-morning or rotating shift schedule — whether on 8-hour, 10-hour or 12-hour shifts. This increase in shiftwork is due partly to society's demands for services around the clock. Grocery stores, pharmacies, catalogue shopping and all types of customer service centers now are open 24 hours a day, seven days a week.

The economics and speed of the global marketplace also are driving companies to expand their hours of operation. Manufacturing plants are running their equipment more, building more products at a lower unit price. Manufacturers and distributors also are delivering products to their customers at a faster pace. Financial services and data-processing companies have to move information on a round-the-clock basis to keep up with the demands of a 24/7 society.

Combined with more traditional shiftwork fields, like fire and police services, medical care, utilities and mining, these new ways of working are increasing the number of people all over the world who work at times other than a Monday-through-Friday day shift.

What Does It Take To Live Well With Shiftwork?

As the number of people working outside of weekday, day-shift hours has increased, scientists and health specialists have learned more and more about how to live and work successfully with shiftwork. These lessons have come from real shift-workers — people just like you who were plagued with serious physical and emotional problems while they worked shifts, and others who seemed to experience little difficulty on shiftwork.

Researchers all over the world have listened to, watched, measured and probed millions of shiftworkers to learn what challenges they faced and why these challenges occurred for some people and not for others. These hard lessons learned from the past are presented here for you, so you do not have to experience the age-old difficulties common to shiftwork.

There are four key factors required to live a healthy, productive life on shiftwork:

▼ *First is the recognition that shiftwork is more than a work schedule. It really is a lifestyle, and it requires new ways of thinking and living.*

▼ *Second is acceptance of the fact that shiftwork adjustment requires an ongoing commitment from you and your family.*

There are no quick fixes, and there never will be any "pills to take away your ills."

▼ *The third key factor in shiftwork success is the support and understanding of family and friends. We all need help to live well with shiftwork.*

▼ *And fourth, achieving* **shiftwork readiness** *— the ability to gain optimal health, performance and well-being, regardless of the time of day or day of the week you work — requires knowledge. Knowing how the body works at different times of the day, how to obtain high-quality sleep, what and when to eat and other important facts are critical to your success on shiftwork.*

The **SHIFTWORK: HOW TO COPE™ Introductory Guide** will provide you with the beginning knowledge you need. Your success with shiftwork is determined by how you use that knowledge. Shiftwork success depends, ultimately, on your intentions — together with the actions you take, starting today!

The Body Clock

So, What Time Of Day Is It Anyway?

Anyone who has worked rotating or fixed night and evening shifts, whether on 8-hour, 10-hour or 12-hour schedules, knows the kinds of challenges that shiftwork can create. The key question is: <u>Why</u> do these problems occur? Knowing the reasons for such difficulties can help us make better choices about how to solve shift-related challenges when we experience them.

Before we start, it's important for you to remember two key points. First, you're not alone with the problems you may be experiencing. Many people all over the world who work night, evening and split shifts have the same troubles. Second, these problems are not a sign of weakness in you or your family. The root source of many of your difficulties is in the design and functioning of the human body.

Humans are daytime animals whose typical life pattern involves sleeping at night and using daytime for work and play. This daily cycle of human activity is no accident. We are all born with internal body clocks, known as the circadian rhythm system, which regulates our body functions. Circadian means "about (circa) a day (dia)."

Our body clocks help us sleep at night by lowering our body temperature, slowing down our metabolic rate and lessening other critical body activities, like digestion, during late-night and early-morning hours. These clocks also control many unseen activities in our body, like the regeneration of cells, kidney functions, the immune system and hormonal secretions.

When our work schedules require us to live in direct conflict with these internal body rhythms — which is the case with anyone who has late-evening, night or early-morning work hours — we may lose track of what time of day it really is. Is it our body's time or the world's time?

For us to perform at our best, both physically and mentally, all of our body functions need to be coordinated or "in sync" with each other. Yet, if we go to bed in the early-morning hours after our evening shift, or have to stay up all night on a night shift, or have to wake up too early in the morning for a so-called "day" shift, then our sleep/wake cycle gets "out of sync" with our other body processes, including our body temperature and digestion. We don't know which internal clock to follow. Even though we might be wide awake at 2:00 a.m., it doesn't mean our body can digest a steak or a piece of pizza.

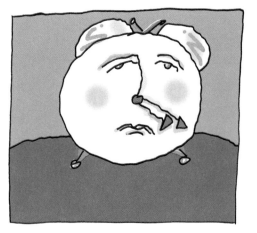

This desychronization of the circadian rhythm system is sometimes called "occupational jet lag," since its effects are similar to those created by traveling across time zones, only more severe and longer lasting. You can reduce or even prevent this "out of sync" experience by following certain circadian principles, especially when planning your sleep, eating and exercise patterns to fit your work schedule.

Why Do Some People Have More Trouble Than Others?

Different people have different types and degrees of difficulty on shiftwork. In fact, one-fifth (20%) of those who work shifts experience only a few, if any, problems adjusting to shiftwork. But 80% of those who have worked shifts in the past reported one or more serious challenges with shiftwork adjustment.

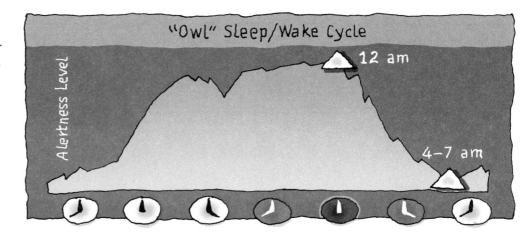

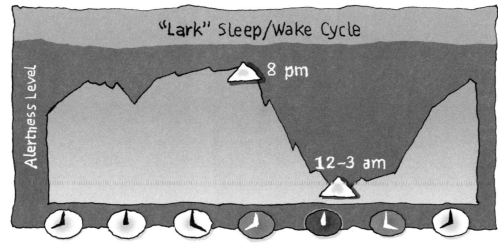

The first factor affecting your **shiftwork readiness** is your chronotype — whether you are a "morning" person or an "evening" person or somewhere in between. Evening-type individuals are known to have fewer problems, at least initially, with an evening or night shift. Their adjustment challenges may occur when they have to get up for a day shift!

The most important factor in shiftwork success, however, is your own attitude and commitment. Are you determined to live a healthy, happy and productive life on shiftwork? Are you ready to learn how to help yourself achieve this **shiftwork readiness** status? If so, then read on and begin to apply these proven shiftwork lifestyle strategies.

Another key factor affecting your ability to tolerate shifts is age. Younger people experience more family and social stress, yet often have more stamina for enduring the physical challenges of shiftwork. On the other hand, older people have a more difficult time with sleep, digestion and fatigue, but may have fewer family pressures to manage.

Sleep

Why Do I Have To Sleep?

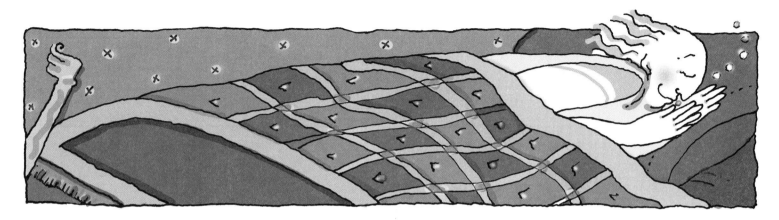

While there still are many mysteries about sleep, there are several key facts that we do know. Most importantly, we know that we cannot perform effectively without sleep. Sleep is essential to proper brain functioning.

Sleep is the time in the day when your brain has the opportunity to restore itself. Sleep is the time for preventive maintenance of the brain. Sleep is not an inactive time. Instead, it is the only time when certain critical brain and body-related activities occur.

Normal sleep is made up of five separate stages. These stages have distinct brain-wave patterns when measured on an EEG machine.

Stage 1 is the drowsy, falling asleep period — the transition between wakefulness and true sleep.

Stage 2 is the light sleep stage during which your brain registers a high degree of alpha waves — the same as the resting state recorded during a period of meditation. You can be easily awakened out of this stage of sleep, as you are "just below the surface" during light sleep.

Stages 3 and 4 are the deep sleep stages when critical mental and physical repair work is done on our brains and bodies. Here, slow waves or delta-wave patterns are evident on an EEG, as all but the most essential brain activities are shut down. Our electrical system, centered in the brain, is being recharged. And the growth hormone is being secreted, stimulating replacement of the many cells of the body. While these deep sleep stages are only about 20% to 25% of a typical 8-hour sleep episode, we cannot function well or for long without deep sleep.

REM sleep is the fifth stage of sleep. During REM, which stands for rapid eye movement, the brain produces beta waves like when we are awake. This is the dream-sleep stage, when our brain is sorting out the day's experiences and entering them into our long-term memory system.

STAGES OF SLEEP

Stage 1

Stage 2

Stages 3 and 4

REM

Well, What's So Bad About Losing Sleep?

When we shorten our sleep length we can become physically clumsy, less able to think clearly and quickly. We also can lose our ability to concentrate and to focus our eyes. Our short-term or recall memory is impaired, and we lose our creativity. We can become irritable, moody, even clinically depressed.

Perhaps the greatest concern, though, is that we can put others at great risk when we are sleep deprived. When we have a poor night's (or day's) sleep, or just do not sleep long enough, it is deep sleep which usually is the first type of sleep that we "lose."

This loss of sleep, known as "sleep debt," sets you up for a microsleep — a brief period (10 to 60 seconds) when the brain drops into deep sleep no matter what you are doing at the time. Microsleep is at the root of many tragic incidents, like the Chernobyl nuclear disaster where 50,000 people died, or the North Carolina traffic accident where 27 children were killed when a truck driver hit a school bus, or the Space Shuttle Challenger accident.

A sleep debt which results in microsleep is the source of most of our accidents on the road, at our workplaces and in our homes. Nothing less than our safety and well-being are at stake.

When Sam first began working a night shift, he found that drinking a few beers after work helped him relax and fall asleep.

It had worked after his day shift, and Sam soon found that it worked just as well after his night shift. Three or four beers made Sam drowsy enough to fall asleep. Sometimes he didn't even bother to eat his morning meal.

Sam thought he was sleeping enough hours, so he couldn't understand why he still felt tired most of the time.

He knew he was eating less, yet he seemed to be putting on a few pounds.

But for Sam, the most important thing was falling asleep, and he had found a way that worked for him. So he stayed with the beer and blamed others for his irritability at work.

A beer or mixed drink seems to help some people to quickly feel relaxed or fall asleep. But does alcohol really help you get restful, restorative sleep? Following are some facts about alcohol and sleep.

BAD NEWS: Most people who use booze to get a snooze will still be tired when they awaken.

QUICK FIX: Beer and wine may help you fall asleep fast, but they won't help you sleep well. Your sleep will be light and disturbed, not deep and steady.

BROKEN DREAMS: Alcohol causes broken sleep. You may not even realize it, but you will awaken often during a sleep brought on by alcohol.

BUSY BLADDER: A few beers before you go to bed also will wake you up often to go to the bathroom, further disturbing your sleep quality.

DISMAL DIET: A person who fills up on beer or other alcoholic drinks doesn't have the appetite to eat well. Alcohol washes away vitamins and gives the body lots of sugar but no food value. It adds to your weight, not to your health.

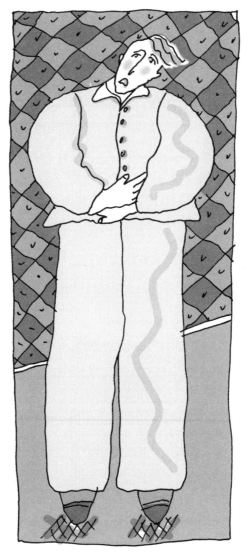

GUZZLER'S GAS: Alcohol irritates the stomach by making it produce more gastric juices. This can cause an upset stomach, diarrhea, and more serious stomach problems. Alcohol can also cause permanent liver damage.

HARMFUL HABIT: It is easy for people who work and sleep at odd hours to fall into a habit of using alcohol, sleeping pills or other drugs to relax or go to sleep. After all, lots of people seem to do it. But many people do not know this simple truth: these substances actually prevent you from getting the very sleep you seek.

Caution! Sleep stealers:

Coffee has become as much a way of life as baseball and apple pie.

Like most adults, you have probably used coffee (or tea or cola drinks) for years as a pick-me-up, a jolt to wake you up and get you going. You think of it as a source of quick energy.

It's the caffeine in coffee that gives you that lift. But it doesn't start working right away. And, caffeine can work just as well to keep you awake when you need to sleep.

Many shiftworkers drink a lot of coffee to stay awake, especially during the long night shift. All that caffeine does the job, but later when it's time to sleep, the caffeine is still at work. It can keep you from falling asleep or, more importantly, it can make your sleep restless. Caffeine disrupts deep sleep.

Caffeine keeps working in the body for about eight hours after it is taken in. So, it's best to stop drinking coffee and other drinks that contain caffeine at least eight hours before you plan to sleep.

Is caffeine stealing YOUR sleep?

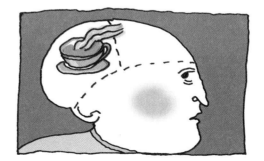

Caffeine hideouts revealed:

Caffeine is found in several popular drinks besides coffee. Following is information about those drinks, and some ideas for snacks that will give you energy without caffeine.

Coffee is a major source of caffeine for most adults. The amount of caffeine in coffee varies with the method used for preparing it.

Caution: Obviously, cutting down on coffee will lower the amount of caffeine you take in each day. But wait — you may be taking in a lot of caffeine every day without knowing it. Below are some of the sources of caffeine that may surprise you.

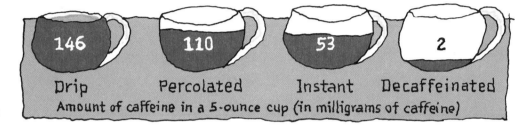

146	110	53	2
Drip	Percolated	Instant	Decaffeinated

Amount of caffeine in a 5-ounce cup (in milligrams of caffeine)

Drinking two to three cans of any of the soft drinks listed below, unless they are labeled "caffeine free," will give you about as much caffeine as you get from one cup of percolated coffee.

Caffeine is found in tea, too, unless it is labeled "decaffeinated." The longer you leave the tea bag in the cup, the more caffeine the tea will contain.

Caffeine is also found in some non-prescription drugs, such as: No Doz, Vivarin, Anacin, Excedrin, Aspirin, Coryban-D, Dristan, Triaminicin, Dexatrim, Dietac, Prolamine, Caffedrine.

Cocoa and chocolate contain small amounts of caffeine, but it would take about 11 cups of cocoa or about six chocolate bars to equal the caffeine in one cup of coffee.

Energy without caffeine:

The natural sugars and vitamins in fruits and fruit juices can give you quick energy without harming your sleep or your health. Try any fresh fruit or some raisins, dates, or figs. Switch from soft drinks or coffee to a fruit juice you enjoy. When you are very tired, drink a small glass of fruit juice every half-hour until you feel refreshed.

Sue doesn't have any problem falling asleep when she gets off the night shift. Her problem is staying asleep.

On Saturday mornings, the children are home and the noise from their playing often wakes her up.

In the summer, the sun beating on her windows makes the bedroom too light and too hot. She wakes up in a sweat and can't go back to sleep.

When she's alone in the house, someone often rings the doorbell or calls on the phone.

Sometimes it's just the noises from outside that wake Sue up — a car's horn honking, a neighbor calling his dog, a church bell ringing — daytime noises that don't disturb people who sleep at night.

People who work evening and night shifts often must sleep when daytime activity is going on around them. And, they must sleep when the sun is up, sending signals to the brain to wake up, no matter how sleepy they might feel.

If your sleep has frequently been disturbed the way Sue's has, you may want to protect your sleep by trying some of the ideas that follow.

Note: *Evening-shift workers, who sleep partly at night and partly in the day, will benefit from these practices, too.*

Strategies For Sleeping

We usually sleep at night when it is cool, dark and quiet. Sleeping during daytime hours can be easier if you fix up your bedroom to create your own night.

▼ Make it dark:

Seal the windows with shades, heavy drapes or aluminum foil when you go to bed, so the room is totally dark.

▼ Make it cool:

Invest in a small window-type air conditioner to keep your room at an even, comfortable temperature. The air conditioner also will give out a steady low hum, which will be relaxing and block out noises from outside.

▼ Make it quiet:

Put heavy rugs or carpeting on the floors to block out noises in the rest of the house.

Keep a low steady sound, like a fan, in your room to block out sudden noises from the outside.

Try to locate your sleeping quarters in an area of the house where there is not a lot of family activity. Don't sleep next to the kitchen or main bathroom.

Educate your family and friends. Children can be taught to be quiet or to play noisy games elsewhere when someone must sleep. Tell your friends when you are working nights so they won't drop in when you are sleeping. If you are alone in the house, you may want to take the

phone off the hook, or hang a "Do Not Disturb" sign on the front door. You can use a pager for emergency calls.

Sometimes something you can't control will disturb your sleep, such as a noisy activity in your neighborhood. Try not to worry about it. Just relax, tell yourself you are safe and let yourself drift back to sleep, minimizing the effect of this waking time on your sleep quality.

Do you have problems getting enough rest when you work the night shift? **Following are some ideas that may help you.** Remember, everyone is different. What works for one person may not work for another. Each human body takes a different amount of time to change sleep patterns.

Try these ideas, one at a time, until you find what works for you. Try each idea for at least a week. If one idea doesn't help, try another.

▼ Relax and wind down:

If you were working the day shift, you wouldn't go home in the afternoon and go right to sleep. You would take time to slow down and relax. When you come home after the night or evening shift, you also need to give yourself some wind-down time. Read, listen to music, write a letter or work on a hobby until you unwind and feel sleepy.

▼ Schedule your wake-up times:

Getting up at a regular time may help you sleep better. When you sleep at odd hours, try to set a regular time to get up, just as you do when you are working days and sleeping nights. Spending too much time in bed may make your sleep shallow. Getting up at a regular time — even if you still feel a bit drowsy — will help you fall asleep more easily and sleep more soundly as the week goes on.

If you should awaken before the time you planned to get up, don't lie in bed and wait for the alarm clock to ring or the appointed time to arrive. Getting up earlier than your scheduled time won't hurt you. If you had a short sleep, just plan to complete your sleep later that day, before returning to work.

▼ Prepare for shift changes:

If you work on a rotating shift, try preparing for your new sleep schedule on your days off before the shift changes.

Whether your next shift is an 8-hour evening or second shift, or a 12-hour night shift, try staying up a little later at night and sleeping later in the morning on your offdays. Easing into the new work-sleep schedule will help your built-in clock adjust to the change faster.

▼ Don't try to force sleep:

You cannot make yourself sleep. If you try to force yourself to sleep when you cannot, you will just become frustrated and more awake. Read, do some gentle stretching, take a warm bath, or do something else relaxing until your body calms down and you feel sleepy.

More Sleep Strategies

When it comes to sleep there is no one just like you. Each person has different sleep needs. Don't worry if your co-workers need more or less sleep than you do. Your only concern should be to get the amount of sleep YOU need to feel fit and refreshed.

Occasionally you will have a day or night when sleep is difficult or impossible no matter what you do. This happens to everyone, not only to night workers or rotating shiftworkers. The problem could be tension at work or at home, an allergy, a virus or "bug", or some other discomfort.

You may feel sleepy and irritable after a sleepless day or night, but losing a few hours of sleep once in a while won't hurt you. It just requires you to take some extra precautions that day.

If the problem continues for a few weeks, however, you might want to see a doctor to find out the cause and perhaps obtain something to help you sleep. It is not healthy to rely on pills to bring on sleep night after night, but using a sleep aid for a few nights of good sleep can break the cycle of insomnia and help you to resume natural sleep.

Remember, though, it takes time for your internal body clocks to change. You should expect some loss of sleep at the beginning of each shift change. Be careful not to confuse this natural sleep loss with insomnia. Sleep should come easier as the week goes on.

There is not one "best" way to sleep. You can, however, adopt habits that will improve the quantity and quality of your sleep.

Following are some ideas that may help you, if you have problems sleeping at odd hours.

Try each idea for a week. If it doesn't help you, keep trying others until you find out what works best for you.

▼ Eat, drink, but be wary:

Hunger can keep you awake or disturb your sleep. Have a light meal or healthy snack before going to bed. Choose foods that are easy to digest. If you eat too much or snack on heavy, fatty foods, your body will be too busy digesting to relax. If your bedtime meal is breakfast, try a bowl of unsugared cereal and milk (unless lactose-intolerant). The milk acts as a natural sleep aid.

Remember, the caffeine in coffee, tea and many soft drinks can keep you awake or interfere with your sleep. Caffeine continues to work in the body for about eight hours after consumed, breaking into your sleep time.

Also, alcohol may help you fall asleep, but it can soon wake you up again or disturb your sleep in other ways. Beer will awaken you faster than liquor, because it increases your need to urinate.

Heavy smoking — about two packs a day — also disturbs sleep, because your body begins to react to withdrawal after three hours without a cigarette. If you give up smoking, you will sleep poorly until your body adjusts; then your sleep will improve greatly.

▼ Set a regular exercise schedule:

Try to exercise regularly every day, even if you are tired. If you keep yourself on a daily exercise program, you will sleep more soundly. Your sleep will improve as your exercise program continues over weeks and months.

▼ Stay active when awake:

If you must stay awake when you are sleepy, activity can help you be more alert. At home, do some outdoor work, take a walk, play a sport. On the job, move around, do some simple stretching exercises, like isometrics. The more active you stay when you are awake, the easier it is to sleep at unusual hours.

Eating

Why Can't I Eat At Any Time Of Day?

Mary had a little lamb,
A chili dog, a chunk of Spam,
Some pizza and a few stewed prunes,
Spaghetti and five macaroons.
She kept loading up her plate;
She dined at six, again at eight.
She cared not what or when she ate
Cause she worked days;
*   she didn't rotate.*

You and your stomach have something in common. Your stomach also works on shifts. The difference is that your stomach does not work night shifts. It is always on the day shift.

Your stomach produces the largest amounts of digestive juices or enzymes during the hours of 10:00 a.m. to 8:00 p.m., when you are expected to be up and eating your main meals. At night, your stomach is basically at rest, sending nutrients into the bloodstream. It is not ready, willing or able to digest much heavy food.

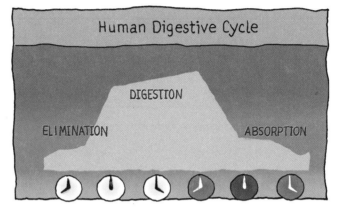

Human Digestive Cycle

DIGESTION

ELIMINATION   ABSORPTION

Put yourself in your stomach's place. You've finished your day's work and are just about ready to do a last-minute cleanup and then go home and get some rest. All of a sudden, someone dumps a whole pile of new food in front of you — a couple of hot dogs with sauerkraut and mustard, a fistful of two-ton steak fries covered with ketchup or maybe globs of chili mixed with raw onions, followed by a big slice of chocolate cake covered with gooey marshmallow and chopped nuts.

Probably, you'd just groan and say, "Forget it. I'm tired. I'll take care of it in the morning."

That's just about what your stomach does. It may make a half-hearted attempt to digest some of that heavy food, but your stomach simply is not prepared to work as hard at night as it does in the daytime.

Don't go to war against your stomach. You will be more comfortable and healthier if you ease up on your stomach's workload by eating foods that are easy to digest when you are working late-evening, night or early-morning shifts.

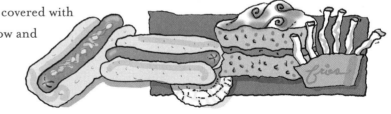

Working on a night or evening shift is something like going into training for a sport. You're making lots of new demands on your body, and you have to treat your body a little differently, too. When you start putting new strain on a muscle in a sport, the muscle lets you know the next day by giving you some pain and stiffness.

Your stomach will let you know, too — with cramps and growls and general discomfort — when you suddenly start loading it up with food at the wrong times. As a result of eating the wrong

foods at the wrong time of day, many fixed and rotating shift employees complain of stomach problems. Some even develop digestive ulcers.

When you eat and what you eat can make a big difference in how you cope with your changing work-sleep schedule. And it will determine the level of nutrition your body receives.

Nutrition is the process by which we take in and use food substances, the way we nourish ourselves.

And proper nutrition is essential to good health. Our body (and brain) need nutrients — like protein, carbo-hydrates, fats, minerals, vitamins and water — to survive and thrive.

Healthy nutrition is about balance. Do not fall prey to food fads or crash programs. As often as possible eat fresh foods, especially fruits and vegetables. And, minimize your intake of fats and sugar.

What Is The Best Eating Pattern For Me?

You need to plan your meals to fit your specific shift schedule and your sleep pattern, while keeping in mind your stomach's primary digestive "window." No matter what shift you work, your main meal(s) should be between 11:00 a.m. and 7:00 p.m., so there is time to complete the digestive process before the enzymes start to decrease for the night.

This also is the time of day you should consume protein foods. Protein is the most difficult food to break down into amino acids, the basic "building blocks" for the body. And, protein requires the most energy to digest. So, eat protein foods during the body's high-energy, high-digestion times.

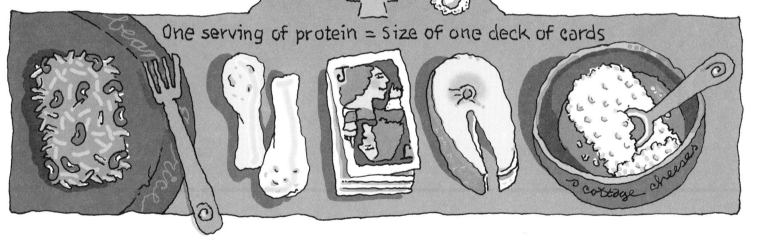

One serving of protein = size of one deck of cards

Meals For The Midnight Muncher: Eating On The Night Shift

Following are some simple rules for eating on the night shift that will keep your stomach from working overtime.

▼ **When you have to eat at night ... think light.**

• *Choose easy-to-digest, high-energy foods such as pasta, vegetables, rice, fruits, whole-grain breads and cereal.*

• *Cut down on heavy, hard-to-digest foods such as red meats, butter, heavy cream, fats, and fried foods.*

• *Avoid spicy foods such as pizza, chili, hot sausage, onions and salsa.*

• *Cut down on caffeine, especially coffee which is highly acidic.*

• *Add fiber to your diet with plenty of fruits, vegetables, whole-grain breads and cereals.*

• *If you eat a meal that is heavy in fat or protein, allow at least three hours of digestion time before heavy work or sleep.*

• *Try to eat one well-planned meal every day. Take time to sit down and enjoy it.*

• *Try to eat at least one meal a day with your family and/or friends. On the evening shift, this might be breakfast.*

▼ *An Upside-Down Meal Plan For The Night Shift:*

Don't worry about following this plan exactly. Just use whatever ideas you think may help you adjust your eating to the night shift.

MAIN MEAL *(4:00 – 6:30 p.m.)*

- **Light Protein** *Chicken, turkey, __OR__ fish (baked or broiled)*
- **Complex Carbohydrates**
  Starches and grains (rice or noodles, lightly buttered);
  AND *Vegetables (yellow or green cooked vegetable; raw vegetable salad or vegetable sticks)*
- **Fruit** *(fresh), frozen fruit yogurt or banana bread*

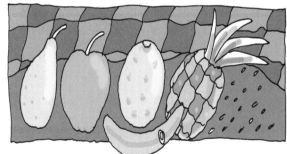

If you are planning to go back to sleep between your main meal and the time you go to work, you may wish to keep this meal even lighter and smaller. In that case, try a bowl of soup or light stew, some whole-grain bread and a vegetable salad.

Avoid red meats, rich gravies and sauces, heavy casseroles or stews, and spicy foods. Main-dish meats, poultry or fish should not be fried.

SNACK *(8:00 – 9:00 p.m.)*

- **Choose one: Fresh fruit** — *apple, orange, banana, pear, pineapple chunks, or other seasonal fruit; raisins, dates, or figs on occasion* __OR__
- **Fruit juice** — *apple, grape, cranberry or other non-citrus juices*

*BEWARE OF CAFFEINE
AFTER 11:00 P.M.!*

"LUNCH" (AT WORK)
(Midnight – 1:30 a.m.)
- **Complex Carbohydrates**
  A cup or bowl of soup (with rice or noodles and vegetables)
  __OR__ A small plate of pasta, lightly covered with mild marinara sauce
  __OR__ A light sandwich on whole-grain bread
- **Fruit** *(fresh)*

Avoid red meats, spicy cold cuts, peanut butter, eggs. Keep this meal light, supplementing with healthy snacks as needed.

SNACK *(3:00 – 4:00 a.m.)*
- **Fruit** *(fresh)* __OR__
  100% fruit juice

BREAKFAST *(7:30 – 9:00 a.m.)*
- **Complex Carbohydrates**
  Grains (like hot or cold cereal; whole-grain toast; or a muffin, lightly buttered)
- **Dairy products** *especially low-fat milk (unless lactose-intolerant)*
- **NO CAFFEINE**

Avoid fried or scrambled eggs, ham, sausage, bacon, or any other fatty or hard-to-digest food; minimize liquid intake before sleep.

Depending on your sleep schedule and the timing of your main meal, you may want to add a light lunch or healthy snack to this meal plan in the early or mid-afternoon.

Hold The Onions, Please! Eating On An Evening Shift

Your first meal of the day should include protein, but remember to keep it lighter and smaller in quantity, especially in the mornings. You only need 6 ounces of protein per day, which is the equivalent of two decks of cards. Surplus amounts of protein put stress on your kidneys and can make you gain weight.

Schedule your exercise so it does not interfere with your digestion. Either exercise before you eat your breakfast, or two to three hours before you consume your main meal of the day.

▼ *Eating Around The Evening Or Second Shift:*

BREAKFAST *(8:00 – 10:00 a.m.)*
- **Light protein** *(such as eggs — soft-boiled, poached or scrambled)*
- **Fruit** *or fruit juice*
- **Complex carbohydrates**
  Grains (like whole-grain toast or a muffin, lightly buttered)

Bacon and sausage contain a lot of fat and are hard to digest, so get your protein from healthier sources.

MAIN MEAL *(1:00 – 2:00 p.m.)*
- **Protein** *(heavy or light; preferably only 3 ounces at a time)*
- **Complex carbohydrates**
  Starches and grains (rice, potatoes or pasta, lightly buttered)
  AND *Vegetables (fresh salad, vegetable sticks or cooked leafy vegetable)*
- **Fruit** *(fresh), frozen fruit yogurt or banana bread*

Red meats, rich gravies and sauces, heavy casseroles or stews, and spicy foods are difficult to digest and can cause discomfort. Main-dish meats, poultry, or fish should not be fried. Try to eat fresh vegetables whenever possible.

"MIDNIGHT SNACK" (OPTIONAL)

(11:00 p.m. – 12:30 a.m.)

- **Milk** and toast

  OR unsugared cereal with milk

  Think light, but you can eat a little to keep from being awakened by hunger pangs in the morning hours. Milk and milk products, especially warmed, can help you relax (unless you are lactose-intolerant).

  Avoid fried foods and anything spicy or hard to digest. This is not a great time for a pizza with everything.

If you plan to go to bed around midnight or 1:00 a.m.,
AVOID CAFFEINATED BEVERAGES AFTER 4:00 OR 5:00 P.M.

"LUNCH" (AT WORK)

(6:00 – 7:00 p.m.)

- **Light protein** (no more than 3 ounces at this time of night)
  Sandwich (chicken or tuna) on whole-grain bread
  OR Cottage cheese or yogurt
- **Complex carbohydrates**
  Vegetables (fresh salad or cooked green or yellow vegetable)
  AND Starches or grains (soup with noodles or rice and vegetables)
- **Fruit** (fresh)

Avoid red meats, spicy cold cuts, peanut butter, and other hard-to-digest foods. Use salad dressing and sandwich spreads sparingly.

SNACK *(8:00 – 9:00 p.m.)*

- **Fruit** (fresh) or 100% fruit juice will give you quick energy, including a lift for the drive home.

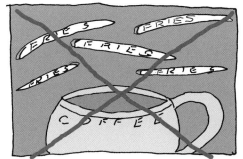

Fatigue

Why Am I Always So Tired?

Fatigue, exhaustion, malaise, wiped out, tired — all of these are words for that feeling we get when we are overextended physically or mentally. We know we feel poorly, but we don't always know exactly what is going on in our bodies.

For some of us, fatigue is temporary. It follows a long day at work or a good workout or an intense, emotional episode. After a little rest, we bounce back and are ready to tackle the next day with renewed energy.

For others, including a high percentage of people on evening, night or rotating shifts, fatigue turns into a chronic condition. We never feel quite rested or restored. Irritability and moodiness set in, creating more tension in our lives and making us feel worse and worse about ourselves. Our bodies feel sluggish, leading us to sit or lie down instead of moving around. The vicious circle has begun, with fatigue breeding more fatigue.

The reasons for high fatigue among shiftworkers are many, including changes in circadian cycles and poor sleep. Fatigue also is linked to physical inactivity, smoking, alcohol use and excess stress.

Sally has been feeling like a stranger to herself lately. She often finds herself avoiding her friends and family.

Little irritations seem to drive her up the wall. She has started smoking and eating more.

She also has stopped doing things she used to enjoy. She finds it hard to relax and have fun. Everything is just too much work!

If you find yourself identifying with Sally, you may be letting stress get the best of you, too. And stress is one of the main contributors to fatigue.

Stress has been recognized as a health hazard; studies have shown that stress can increase the risk of heart disease and have harmful effects on job performance and quality of life.

Working on a schedule of evening, night or rotating shifts can increase the amount of stress you have in your life. You operate on what seems to be your own personal schedule, out of step with the rest of the world. When you are awake, everyone else is sleeping. When you must sleep, others are going places and doing things.

You eat on a different meal schedule. Sometimes you miss a meal entirely. Sometimes you can't sleep, and it seems as though you will never feel really rested again.

And then, there are all the stresses of daily life as well — money problems, car troubles, family fights. Sometimes it seems as though you just can't win. But you can. The key is to learn to control stress instead of letting it control you.

If stress is running your life, you may want to try some of the suggestions which follow to help turn the situation around.

How Can I Relax And Restore My Energy?

The more comfortable you feel emotionally, the better you will be able to handle the day-to-day stresses in your life. Some of these ideas may help you ease your mind, so you can relax and feel rested.

▼ *Give yourself a break:*
Some simple relaxation exercises can help you handle stress. Try one of the exercises outlined on pages 40 and 41. The first two can be done anywhere, any time you need help to calm down.

Join a club, church, athletic team, or some other organization whose activities interest you. Don't worry about being the most active member. Participate when your work schedule will allow you to, as long as you find the activity relaxing.

Save some quiet time all to yourself every day. At least once a week, try to spend time doing something just for fun!

Try to spend some time each day with your spouse, or family, or close friends. It's not how much time you spend, but what you do with the time that's important. *You may find more companionship in a ten-minute walk with someone than in a whole evening watching television together.*

▼ Talk it out:

Sharing your burdens is a good way to reduce stress. When something is worrying you or upsetting you, don't just keep it inside and dwell on it. Talk it over with your spouse or a close friend.

Too many decisions? Try sharing those, too. Decisions about money, chores, work schedules, and daily living issues are easier to handle if you share them with people you work with or live with.

▼ Hang loose:

When you are faced with a new idea or experience, give it a chance. Think about it or try it out for a day or so before making up your mind one way or another.

Some shiftworkers say they feel guilty if they must sleep when their family would usually be together. If this is a worry for you, try to remember that your work and your health are important to your family's well-being. If you are sleep-deprived or fatigued, you are of no use to yourself — or anyone else!

▼ Check your time:

Don't try to cram too much into your time off work. Be realistic about how much you can do in the time you have available.

Try to keep a handle on where your money is going. Piling up a lot of bills can create additional worries.

Relaxation Exercises

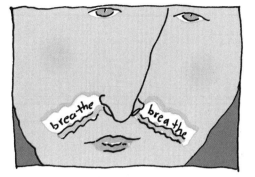

▼ **Take a deep breath:**

Smooth, slow breathing can help you control anger or other strong feelings. It is important to breathe from the diaphragm and abdomen. If your stomach goes up and down when you breathe, you are breathing properly. Breathing through the mouth or through the shoulders increases tensions.

1. *Take smooth, slow breaths through your nose. Breathe out slowly until you are taking twice as long to breathe out as you are to breathe in.*

2. *Count to three as you breathe in. Count to six as you breathe out.*

3. *Continue this breathing exercise for a few seconds or minutes until you feel the tension begin to leave your body.*

▼ **Magic pictures:**

1. *Picture a calm scene in your mind — a leaf floating on a pond, clouds drifting across the sky, or something else that is pleasing to you.*

2. *Imagine this scene in the greatest detail possible. Your muscles will relax as you concentrate.*

▼ Total relaxation:

For best results, practice this exercise once or twice a day, 15 to 20 minutes each time, but preferably not right after a meal.

1. *Find a quiet spot. Sit comfortably or lie on your back. Uncross your arms and legs. Close your eyes.*

2. *Relax your muscles, beginning with your feet and working up to your face. You may find it helpful to tense each muscle first and then relax it until it feels warm and heavy.*

3. *Clear your mind of thoughts and concentrate on your breathing. Breathe slowly through your nose, drawing deep breaths from your diaphragm. As you breathe out, say a one-syllable word or sound to yourself such as "one" or "um" or "slow."*

4. *If distracting thoughts drift through your mind, don't concentrate on them. If a thought comes back, say "no" or "go" under your breath.*

5. *You may open your eyes to check the time, but don't use an alarm clock. When you have finished the exercise, sit quietly for a few moments with your eyes open, and then resume your regular activity.*

What Is "The Fitness Factor"?

The better shape you are in physically, the better you can handle day-to-day stresses. One of the best ways to stand up to stress is to meet it in good physical health.

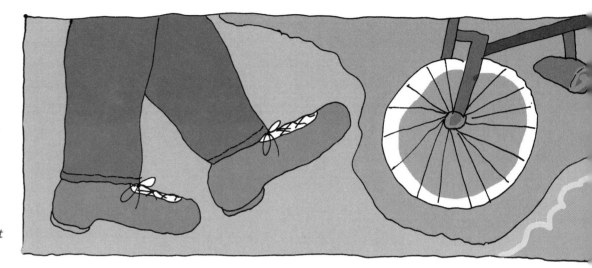

▼ Move some muscles:

Set up a program of regular exercise or activity for yourself — walking, biking, jogging, swimming — whatever appeals to you. Try to do your chosen exercise at least three times a week, for 30 minutes at a time. Start out with the amount of effort you can handle comfortably and build up gradually. Exercise should be fun, not pain and misery.

Slowly increase your level of exercise to where you reach the point of perspiring at least twice a week. This will increase your heart rate and strengthen your heart muscle.

Also, do some sort of mild muscle-stretching exercise daily. When you wake up or before you go to bed are both good times of the day for stretching.

or cut out your use of anything that can harm your health.

Limit coffee and other drinks that contain caffeine. Caffeine will "rev you up," but it also will increase your tension level and lead to poor sleep.

Set some limits on the amount of alcohol you have each day and each week. A reasonable goal might be no more than two beers or one mixed drink on an every-other-day basis.

If you can't stop smoking entirely, limit yourself to a certain number of cigarettes each day. Then, if you want to continue cutting down, gradually smoke less of each cigarette.

▼ Build from the inside:

Treat yourself to at least one well-planned meal each day. Take time to sit down, relax and enjoy it.

If extra pounds are dragging you down, get rid of them slowly but surely. You will feel better physically and mentally if you bring your weight down to what's right for your height. Looking in the mirror can be fun again.

Think about what you're putting into your body when you eat, and cut down

On-Shift Alertness

How Can I "Rev Up My Engine" At Work?

Human alertness levels vary with the time of day. When you work during the low points of this alertness cycle — either the "post-lunch dip" time in the afternoon or the "early-morning trough" time from 1:00 a.m. to 5:00 a.m. — you have to have some ways to counter those drops in alertness.

When alertness drops, attention drifts. You are likely to make careless errors or miss things entirely. And, sometimes, these mistakes can have serious consequences.

Besides the time of day on your circadian or body clock, there are five primary factors that control alertness. Attending to each of these will help you to stay safe and work productively on your shift.

On-Shift Alertness Strategies

▼ *Make a daily "deposit" in the sleep bank:*
*Loss of sleep or sleep deprivation creates a sleep "debt." The only way to prevent this debt from accumulating is to get the sleep your brain requires every 24 hours. As the Sleep section of this **Guide** makes clear, you need sufficient amounts of deep sleep and REM sleep every day in order to be fully awake and alert. When you come to work with a sleep "debt," you put yourself and others at risk.*

▼ Stay motivated:

A high level of interest and focus can help you to counter low alertness. So, during these critical times of the day, you need to pay attention to what you are doing. This is no time to take shortcuts or to let your attention wander.

▼ Keep moving:

Muscle activity will reduce muscular fatigue and boost alertness. Even small-motor activity, like isometric exercises or a change in posture when you are sitting, will help to re-stimulate you. Large-motor activities, like walking or a short stint of calisthenics, also will trigger a "fight-or-flight" response from your sympathetic nervous system. Aside from sleep, physical activity is the best way to increase alertness.

▼ Seek bright light:

The brain is awakened by exposure to bright light. So, make sure your work environment is well-lit. Don't be tempted to turn the lights off in order to see your computer screen better. Remember, when you turn off the lights, you also turn off your brain.

Many of us are in the habit of thinking about a break just as a time to stop work. We often go to a break room and do one or more of the following things — sit down, consume caffeine, smoke cigarettes and/or complain! And all four of these activities are de-energizing.

So, it is no wonder we feel tired and have low alertness at work. In large part, we are doing this to ourselves! Instead, **we need to think of a break as a time to restore ourselves**.

▼ *Use break time wisely:*
- *Drink fruit juice.*
- *Eat fresh fruit.*
- *Move around on break.*
- *Have a positive, stimulating conversation with someone.*

With these practices, we will be more alert and leave work with energy leftover for ourselves.

Many people working shifts express concern about driving home from work when they are tired or sleepy or both. To be sure you are safe on the road, **here are some actions you can take to raise your alertness level.**

Driving Strategies

▼ *Drink fruit juice*

about 20 to 30 minutes before you get in the car to drive home. The "good" sugar (fructose) found in 100% fruit juice will give you a quick lift.

▼ *Move before you go*

especially if your job involves sitting for all or most of the day. Even if you move around on the job, some quick toe-touching or running in place or even some isometric stretching can wake you up for the drive home.

▼ *Vary the route*

that you travel. If your brain has to think about where you are going, it is less likely to drop into a microsleep.

▼ *Take a passenger*

or travel in a carpool so you will have someone to talk to and someone to watch out for you. Take turns driving if you begin to feel sleepy or notice your energy dropping.

▼ *Get fresh air*

or at least cool air moving in the car, by opening your vents or a window (if it is safe to do so). Oxygen flowing to the brain will help raise your alertness level.

Remember that you cannot simply tell yourself to stay awake if you are sleep deprived and your brain wants to take a microsleep. Take these precautions before you get behind the wheel — your life or someone else's may depend on it!

Family And Social Time
(the rest of your life!)

Why Doesn't Anyone Understand Me?

Shiftworkers often feel alone, unappreciated and misunderstood. Children complain about their parents being grumpy. Spouses complain that they are left alone to shoulder all of the household responsibilities.

If these concerns and complaints are left to fester, family distress and even separation and divorce can result. Also, people working rotating or fixed shifts can find themselves in a state of social isolation, cut off from friends and social activities.

Changing this situation — or preventing it from happening in the first place — requires everyone in the shiftworker's household and friendship network to learn more about shiftwork. When we educate our family and friends, we let them into our world. We give them a chance to "walk in our shoes."

And we open up the lines of communication. Talking and sharing are the first steps to understanding.

Jim found his odd free-time hours one of the biggest hassles of working rotating shifts.

Being an outdoor person, he liked having daytime hours free when he worked evenings and nights. But he often missed spending time with his family, getting together with his friends, or joining in team sports.

The evening shift seemed the most difficult to Jim because his free time was split into a few daytime hours and a few late-night hours. He found it hard to work on household projects and still have time to relax before work.

Once Jim tried to paint his garage while he was working the evening shift. The time was so limited that he found himself rushing to work every day tired and tense after a last-minute cleanup. He was grouchy to others on his shift and snapped at his wife when he got home.

Usually after the evening shift, Jim found it relaxing to spend some quiet time with his wife, do a few stretching exercises, have a late snack and watch some television before going to bed.

During his painting project, Jim was in no mood to talk to anyone when he got home. It irritated him even more to think that, on top of everything else, he hadn't had a chance to see his children all day.

He put off his exercise, substituted a six-pack for the snack, and woke up still feeling grouchy the next morning.

Work is important. It should give you a sense of pride and accomplishment. It is an important part of the quality of your life.

But work should not be your whole life, and your time off should be purposeful and pleasurable, not just a breather from the job or a time to do more work.

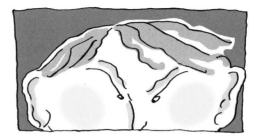

Here are some ideas for planning your time that may help you avoid the frustrating situation Jim set up for himself. Some planning, and imagination, may give you time for more pleasure in your time off.

Time Management Strategies

Planning your free time may at first seem like just another burden to deal with, but if your time becomes more pleasurable as a result, then planning will become easier.

To get the most from your free time, you might set up your plan in the following way:

▼ **Plan what you MUST do:**

First, plan time for the essential tasks you MUST complete that day. Be realistic about how much you can fit into the time available and what is the most important to you.

If you are planning a major household project, determine if and how the work can be spread over the free time you have on the shift you are working. Ask yourself if you are planning the project in the best phase of your shift schedule. A project that could cause a time crunch on one shift might be better left until you are working on a different shift.

▼ Plan what you WANT to do:

Be sure to set aside specific times in your schedule to relax, eat, exercise, visit a friend, or whatever activities you like to do for pleasure.

▼ Stick to the schedule:

Don't procrastinate. Putting things off is the best way to end up frustrated and tense. Halfway through your schedule, stop and see how you are doing on time. If there is too little time left to finish all you had planned, decide again what are the most important tasks to complete and do only those.

▼ Plan for some overflow time:

Include extra time in your schedule for interruptions, equipment breakdowns, and other delays. That way you won't find yourself rushing frantically at the last minute.

▼ Use your imagination:

Try to be creative when there is something you really want to do that your shift schedule won't allow. The challenge of arranging an activity may make it more enjoyable.

Sally hated to give up her weekly bowling, but her new night-shift assignment made it impossible for her to bowl with her regular team. She talked with other people on her shift and found enough to make up a new bowling team. This team planned its bowling activities to fit their night-shift schedule.

Bill found he was not seeing his children at all when he was on the evening shift. They left for school before he got up and came home after he had gone to work. He volunteered to help out occasionally at his children's school, as a noontime playground supervisor, when his work schedule allowed. This not only gave him time with his children but a chance to get to know their teachers and classmates, too.

Jim's favorite sport is baseball. When his favorite team made it to the World Series, Jim found that most of the games would be televised while he was at work. He asked a friend to tape the games for him, rented a VCR, and enjoyed watching them when he got home from work.

Lisa missed her small children when she was on the evening shift and could not have supper with her family. Once a month during her evening shift she packed a picnic supper before going to work. Then her husband and children would join her on her lunch break and have supper together in the employees' lunchroom. This also gave Lisa's children a chance to see their mother's workplace.

John's 12-hour night schedule left him little time for any regular exercise or social activities. All of his leisure time was lumped into his days off, while his workdays seemed to be spent on just work and sleep. Recently, he discovered how to get his daily dose of social time and stay physically fit, by planning a one-hour basketball game in the afternoon with friends from work. This scheduled workout still left him with sufficient sleep and meal time. And also it gave him important social time with his friends every day.

Family Matters!

Over and over, shiftworkers have said that support and understanding at home are the most important aspects of adjusting successfully to shiftwork.

Shiftwork affects the whole family, so an important consideration for any shiftworker must be, "Will my family be able to handle my schedule?"

The problems differ for each family, depending on: the makeup of the family, the ages of your spouse, your children (if you have any) and you; your interests as individuals and as a family; and, above all, your ability to communicate with each other.

Some of the decisions you must make are:

If as a couple you can adjust your sex life and your social life to your unusual working hours.

If your family can handle problems alone that might arise while you are at work.

If your family can cope with your extended absence when your relief fails to show up and you must work an extra shift.

If you and your family can feel comfortable when they are alone during your evening and/or night shifts.

If your spouse can cope with the burden of making many child-raising and household decisions alone because you are at work or asleep. If your spouse and children can understand and respect your need to sleep during times when they are active.

If your family can willingly assume some share of the household duties that your schedule may prevent you from performing.

If the extra money usually paid for night or rotating shiftwork is worth the sacrifice of time together that your family must make.

If your family is understanding enough to cope with the tenseness and irritability you may experience as a result of your work hours.

None of these challenges may happen to you and your family, but it is important to address the worst-case scenario while planning for the positive benefits that shiftwork can create.

Communication Strategies

Once you have made the decision to work an evening, night or weekend shift, it is important to stay alert to how shiftwork is affecting you and your family.

Following are some ideas for ways you and your family might be aware of changes that are happening in your lives as a result of shiftwork.

▼ *Keep talking:*

Communication between you and your spouse is key to how you and your family cope with shiftwork. Most problems can be dealt with if they are picked up and talked over before they grow so big that they get out of hand. Try to set aside some regular time each week to talk with your spouse about what is going on in your lives as a result of your work hours.

▼ *Lend an ear:*

Try to be aware of what is going on in your children's lives as a result of your work hours. Helping them to understand the difficulties of your work schedule may make their adjustment easier.

▼ Quality counts!

The way you spend time with your family can often make up for the time you miss together. Taking a son or daughter alone with you on special trips, such as a sports event or a shopping trip, will mean more to you and your child than a few nights together in front of the television set.

▼ Plan together:

Decide together what activities you do not wish to give up as a family, and see how you can fit those activities into your schedule. When your work schedule conflicts with a special holiday, for example, you and your family may be able to plan a family celebration around your schedule. If a family event is going to occur during your work hours, decide if the event is important enough for you to take time off, remembering that later you might have to work an extra shift to make up for that time.

Consultants And Contributors

Sara M. Felch, R.D.
  Nutrition Department
  Upper Valley Medical Center

Michael S. Gaylor, M.D.
  Behavioral Medicine
  Dartmouth Medical School

Peter Hauri, Ph.D.
  Sleep Physiology
  The Mayo Clinic

Eugene C. Nelson, D.Sc.
  Health Education
  Dartmouth Medical School

Olov Ostberg, Ph.D.
  Ergonomics
  National Institute of Occupational
  Safety and Health

Ellen Roberts, M.P.H.
  Health Education
  Dartmouth Medical School

Michael A. Schneider, Psy.D.
  Clinical Psychology
  Counseling Center
  of Sullivan County

Jeannette J. Simmons, D.Sc.
  Health Education
  Dartmouth Medical School

Susan C. Linsey, M.S.
  Research Associate
  Dartmouth Medical School

Georgia Croft
  Writing Associate
  The Valley News

Illustrations:
  Dale Gottleib and
  Nancy Montgomery

Book Design and Production:
  S.T. Vreeland Marketing & Design
  Yarmouth, Maine
  Nancy Montgomery, designer